SAFETY-VALVES.

BY

RICHARD H. BUEL, C. E.

(REPRINTED FROM THE "RAILROAD GAZETTE.")

NEW YORK:
D. VAN NOSTRAND, PUBLISHER,
23 MURRAY AND 27 WARREN STREET.
1875.

INTRODUCTION.

The writer, in presenting these remarks to engineers, does not pretend to offer much that is original, but has aimed to gather what is valuable from the great mass of material to be found in scientific periodicals and in publications that are not generally accessible. An endeavor has been made to systematize the treatment of the subject, and to give such varied solutions of the problems that arise in proportioning the parts of safety-valves as to render them plain to those who have only an elementary education. The importance of having the general principles of safety-valves understood by those who are charged with the care of steam machinery cannot well be overestimated. With a safety-valve that is in reality all which its name implies, a large proportion of the risks incident to the use of boilers will be avoided; while on the other hand, a safety-valve that is only such in name is one of the readiest assistants to a disastrous boiler explosion.

NEW YORK, August, 1875.

SAFETY-VALVES.

I. The Requisite Qualifications of a Safety-Valve.

As a safety-valve is designed to prevent the accumulation of pressure in a steam boiler beyond a certain point, it is necessary that the parts should be so proportioned that the valve will rise when the given pressure is attained. Until the valve rises it is subjected to the pressure of the steam at rest, so that this part of the subject involves the statical condition of a safety-valve. As soon as the pressure of the steam in a boiler lifts the valve, new conditions are introduced, because the steam is in motion, escaping through the orifice between the valve and the seat. It will thus be evident that from the time the valve is raised until it is again seated by the reduction of the steam pressure the dynamical conditions are to be regarded. A good safety-valve should be so constructed that not only will it lift when the required pressure is attained, but so that it will also prevent the further increase of pressure, and will close promptly as soon as that pressure is reduced.

II. Proportioning the Parts of Safety-Valves, in order that they may Rise with Given Pressures.

This part of the subject, as already remarked, deals with the statical condition of safety-valves. In other words, it is a question of the equilibrium of two forces acting in contrary directions—one, a weight or the tension of a spring, tending to hold the valve down; and the other, the pressure of the steam, tending to raise it. When these opposing forces balance each other the valve is ready to lift, and any slight increment of pressure will raise it. It is, then, the conditions of the balancing or equilibrium of the steam pressure and the spring or weight that are to be considered. In some forms of safety-valves, a spring or weight is placed directly above the valve, and resists the upward pressure of the steam; or a weight is suspended directly under the valve, passing into the boiler. In other forms, the spring or weight is attached to a lever, to which the valve is also connected, the weight or spring being ordinarily at a greater distance from the fulcrum than the valve is. In the first form of construction, in which the steam pressure is opposed directly by the force of a weight or spring, without the intervention of a lever, these two forces will evidently balance when they are equal to each other. It is only necessary, therefore, to multiply the pressure of the steam in pounds per square inch by the area of the valve in square inches to find what weight must be attached, or what tension put upon the spring, to prevent the valve from rising before this pressure is reached. For example, suppose that a valve has a diameter of 4 inches, and is required to rise when the steam pressure is 100 pounds per square inch, what weight must be attached to it?

The area of a valve having a diameter
of four inches is about..12 9-16 square inches.
Multiply by........... 100

Weight to be attached to
valve....... ....1,256¼ pounds.

In the second case, where a lever is employed, it is evident that the forces will not balance if they are equal, since they act at different points of the lever. Fig. 1 is a sketch of an ordinary lever

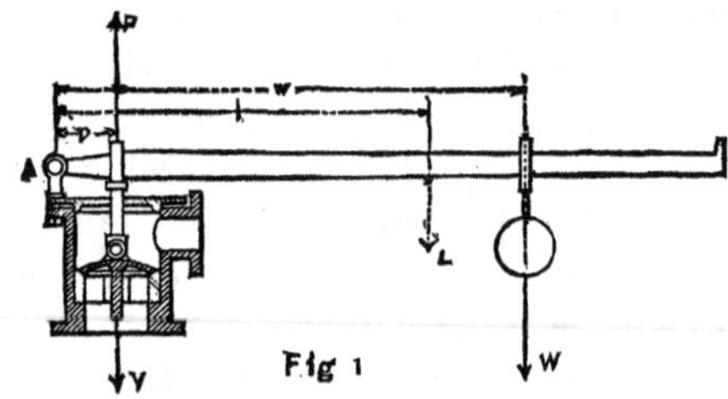

Fig 1

safety-valve, and the forces acting in a

case of this kind are represented graphically by arrows, the directions in which they point corresponding to the directions in which the forces act. It will be seen that there are four forces :

1. The weight, represented by the arrow W.

2. The weight of the lever, represented by the arrow L.

3. The weight of the valve and stem, represented by the arrow V.

4. The pressure of the steam, represented by the arrow P.

The latter force, the pressure of the steam, is the pressure per square inch multiplied by the area of the valve in square inches. The lever is arranged so that it can turn about the point A as a fulcrum. The first three forces tend to keep the valve down, the fourth force tends to raise it, and all the forces act vertically. The lever arm of a force is the distance from the force to the fulcrum, *measured on a line that is drawn perpendicular to the direction of the force.* As these forces act vertically, the lever arms must be measured on horizontal lines—and it will be seen that there are three lever arms to be considered :

1. The lever arm of the weight, represented by w.

2. The lever arm of the lever, represented by l.

3. The lever arm of the valve, represented by p.

The lever arms of these forces are the horizontal distances from the centers of gravity of the weight, lever, and valve, respectively, to the fulcrum. The centers of gravity of the weight and valve are ordinarily in vertical lines passing through their centers, and the center of gravity of the lever is most readily determined by balancing it upon a knife edge, the center of gravity being on a vertical line passing through the point on which it balances.

To recapitulate, there are eight quantities that may be varied, in proportioning a safety-valve, viz.:

1. The weight.

2. The weight of the lever.

3. The weight of the valve.

4. The diameter of the valve.

5. The lever arm of the weight, or its position on the lever.

6. The lever arm of the lever, depending upon its form.

7. The lever arm of the valve.

8. The pressure of the steam, in pounds per square inch.

Any seven of these parts may be proportioned at pleasure, and the remaining one can be determined, the sole condition being that the valve shall rise as soon as a given pressure is exceeded. Ordinarily, however, there are only two cases that require solution :

1*st. With what steam pressure will a given valve rise?* 2*d. Where must the weight be placed on the lever of a given safety-valve in order that the valve may rise when a certain steam pressure is reached?* These problems can be solved either by experiment or calculation. Both methods will be explained.

A.—Experimental Method.

Ascertain the weights of the ball, lever and valve, the diameter of the valve, the center of gravity of the lever, and the lever arms of the weight, lever and valve. The lever arms of the weight and valve can be measured directly, and the lever arm of the lever, or the horizontal distance from its center of gravity to the fulcrum, can be determined by balancing it upon a knife-edge, as already explained. In measuring the valve, if the seat is beveled and the valve fits well, the smallest diameter is to be taken. Then provide a bar, *A B*, Fig. 2, of uniform section, at least twice as long as the distance from the weight to the fulcrum of the given valve. Mark the center-point, *C*, of the lever, and points *E*, *F*, *G*, to the right of the center, such that the distance *C E* is the same as the horizontal distance of the center of the valve from the fulcrum, *C F* is the horizontal distance from the center of gravity of the lever to the fulcrum, and *C G* the horizontal distance from the center of

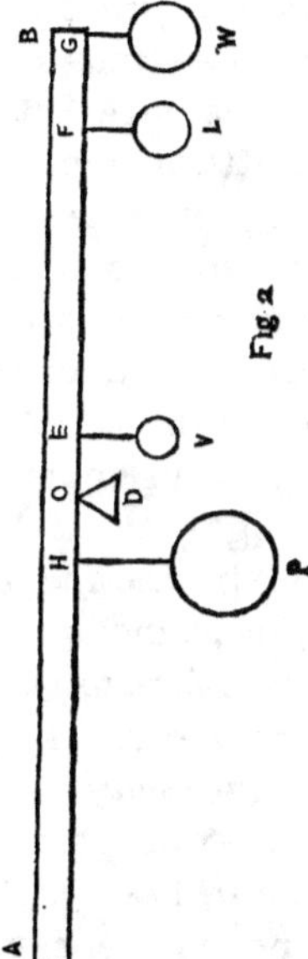

Fig 2

the weight to the fulcrum, if this is known. Also lay off to the left of the center a distance, $C\,H$, equal to $C\,E$. Balance the bar upon a knife-edge, D, and the apparatus will be ready for making the required determinations.

1st. *To find with what steam pressure the valve will rise.*—Hang on the ball at the point *G*, the lever at *F*, and the valve at *E*, or attach equivalent weights, as may be most convenient. Then hang on weights at the point *H*, until they just balance the weights on the other side of the center, and bring the bar to a horizontal position. Ascertain the area of the valve in square inches, and divide the weight at *H* by this area. The result will be the required pressure of the steam, in pounds per square inch.

2nd. *To find where to place the weight on the lever of a given safety-valve, in order that the valve may rise with a given steam pressure.*—Multiply the area of the valve in square inches by the pressure of the steam in pounds per square inch, and attach a weight at *H* equal to this product. Also hang on the valve and lever, or equivalent weights, at *E* and *F*, as before, and move the ball along the bar until it balances the other weights. The distance from *C*, of the point at which it balances will be the horizontal distance from the fulcrum to the weight, on the lever of the given safety-valve.

To facilitate the only calculation required in this experimental method, the areas of valves are given below, for the majority of cases that occur in practice :

Table of Areas of Valves of Different Diametres.

Diameter of Valve in inches.	Area of Valve in square inches.
½ or 0.5	13-64 or 0.19635
⅝ or 0.625	5-16 or 0.30680
¾ or 0.75	7-16 or 0.44179
⅞ or 0.875	19-32 or 0.60132
1	25-32 or 0.7854
1¼ or 1.25	1 15-64 or 1.2272
1½ or 1.5	1 49-64 or 1.7671
1¾ or 1.75	2 13-32 or 2.4053
2	3 9-64 or 3.1416
2½ or 2.5	4 29-32 or 4.9087
3	7 1-16 or 7.0686
3½ or 3.5	9 5-8 or 9.6211
4	12 9-16 or 12.5664
4½ or 4.5	15 29-32 or 15.9043
5	19 41-64 or 19.635
5½ or 5.5	23 49-64 or 23.7583
6	28 9-32 or 28.2744

The method here described is exceedingly simple and accurate, and will, it is hoped, find favor with those who have difficulty in using the ordinary modes of calculation. It is not uncommon for the graduations on many safety-valves to be wrongly marked, and it is important for steam users to know certainly whether or not they are correct.

A simple experimental test of the accuracy of the adjustment of a safety-valve may be made without removing it from its position on the boiler. The method is described below, the account being taken from an article by the writer published in the *Scientific American* for Oct. 31, 1874.

"Secure the valve stem of the safety-valve to the lever with wire or string, and attach a loop to the lever, into which pass the hook of an accurate spring balance, arranging the loop so that it is directly over the center of the valve stem. Then take hold of the upper part of the spring balance and lift the valve slightly, noting the reading of the balance. Measure the lower diameter of the safety-valve, and find its area; divide the reading of the spring balance by the area of the valve, and the result will be the pressure, in pounds per square inch, at which the steam will raise the safety-valve. Suppose, for instance, that the diameter of the safety-valve is 1 inch; its area will be about $\frac{7854}{10000}$ of an inch. Now, if the tension of the spring balance in raising the valve is 120 pounds, the pressure at which the valve will rise is the quotient arising from dividing 120 by $\frac{7854}{10000}$, or 153 pounds per square inch."

B.—Graphical Method.

The statical problems in connection with safety-valves can also be solved graphically or by construction. A force can be represented in intensity and direction by a straight line, by assuming some unit of length for the unit of intensity. For instance, if the force is 5 pounds, it may be represented by a line $\frac{5}{8}$ of an inch in length. To simplify the graphical solution, the several constructions upon which it depends will be presented separately.

1. *Two or more parallel forces, acting in the same direction on a lever, being given, to find the direction, intensity, and point of application of a single force which can replace them and produce the same effect.*

In Fig. 3, the lever, *A B*, is acted upon by two parallel forces, *C D*, and *E F*. Lay off horizontally to the right of *E*, and the left of *C*, any two distances, *E a*, *C b*, equal to each other. From *a* draw *a c* parallel to *E F*, and from *F* draw *F c* parallel to *E a*, forming a rectangle, *E a F c*, on the two adjacent sides, *E a*, *E F*. Also complete the rectangle, *C b d D*, of which *C b*, *C D*, are the adjacent sides. Through the points *C* and *E* draw diagonals, *d C*, *c E*, to the two rectangles, and produce them until they meet in the point, *e*. Through *e* draw a line, *e H*, parallel to *C D* and *E F*, and the point *G*, in which it cuts the lower line of the lever, will be the

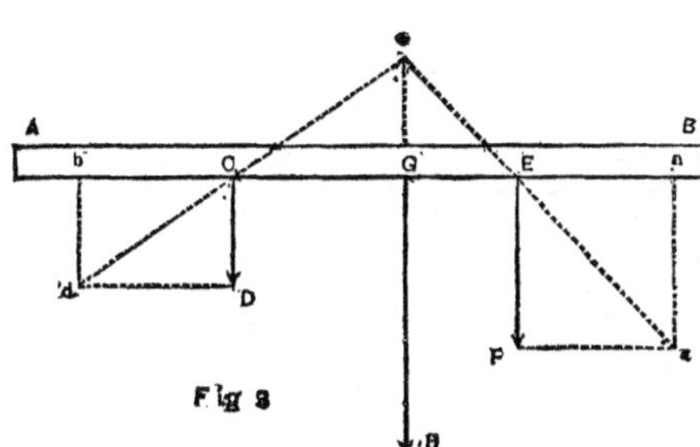

point of application of a force which will singly produce the same effect as the two forces, CD, EF. Make GH equal to the sum of CD, and EF, and it will represent the intensity of the required force, which is called the resultant of the other two forces.

If there are three forces to be replaced by a single one, after finding the force that is equivalent to two of them, consider this and the third force as two single forces, and proceed as before. In this way any number of parallel forces acting in the same direction can be replaced by their resultant, which is a single force, equivalent in effect to all the others.

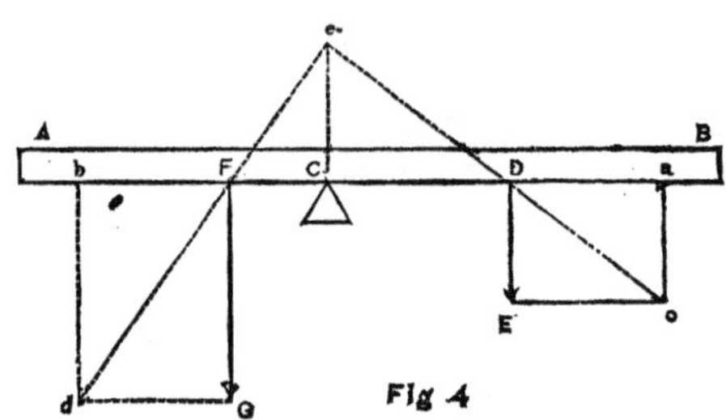

Fig 4

2. *The intensity and point of application of a vertical force acting on a lever being given, also the position of the fulcrum, to find the intensity of a parallel force, acting at a given point of the lever, which will balance the first force.*

D E, Fig. 4, is the given force, *C* the fulcrum, and *F* the point at which the required force must act and balance *D E*. Draw a vertical line, *C e*, from the fulcrum. Lay off, to the right of *D* and the left of *F*, any two distances, *D a*, *F b*, equal to each other. Complete the rectangle, *D a c E*, and draw a vertical line, *b d*, from the point *b*. Through *D*, draw a diagonal to the rectangle *D a cE*, and produce it until it meets the vertical line through the fulcrum in the point *e*. From *e*, draw a line through *F*, the point of application of the required force, continuing it until it meets the vertical line drawn from *b* in the point *d*. Complete the rectangle of which *b F*, *b d*, are the adjacent sides, and *F G* will be the force required to balance *D E*.

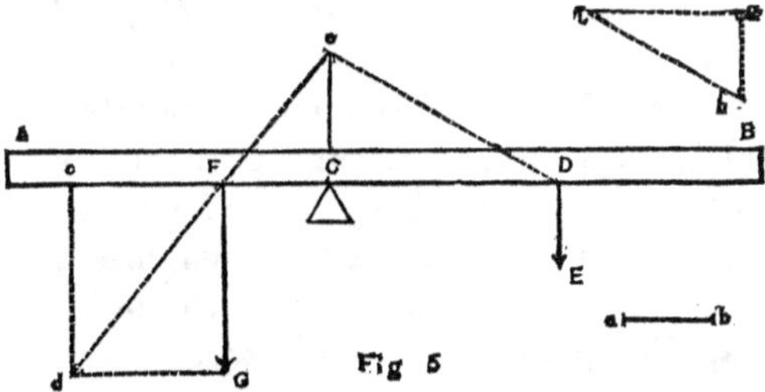

Fig 5

3. *The intensity and point of application of one vertical force, the intensity of a second parallel force, and the position of the fulcrum, being given, to find the point of application of the second force, in order that it may balance the first one.*

F G, Fig. 5, is the force whose intensity and point of application are given. *C* is the fulcrum, and *a b* is the intensity of the second force. Draw a vertical line, *Ce*, from the fulcrum. Lay off from *F*, towards the left, any distance, *F c*, and complete the rectangle *F c d G*. Through *F* draw a diagonal to the rectangle, continuing it to *e*, where it meets the vertical line drawn from the fulcrum. In any convenient place, lay off a horizontal line, *f g*, equal to *F c*, and from the point *g* draw the perpendicular *g h*, equal to *a b*. Through the points *f* and *h*, draw the line *f h*, and from *e* draw a line, *e D*, parallel to *f h*. The point *D*, in which the line last drawn meets the lower side of the lever, is the point of application of the second force ; and *D E*, equal to *a b*, represents that force applied at such a point as to balance *F G*.

By the application of the graphical methods just explained, it will be easy to solve the problems previously determined by experiment.

1*st*. *To find with what steam pressure a safety valve will rise.*—On the lever *A B*, Fig. 6, mark the fulcrum *C*. To the

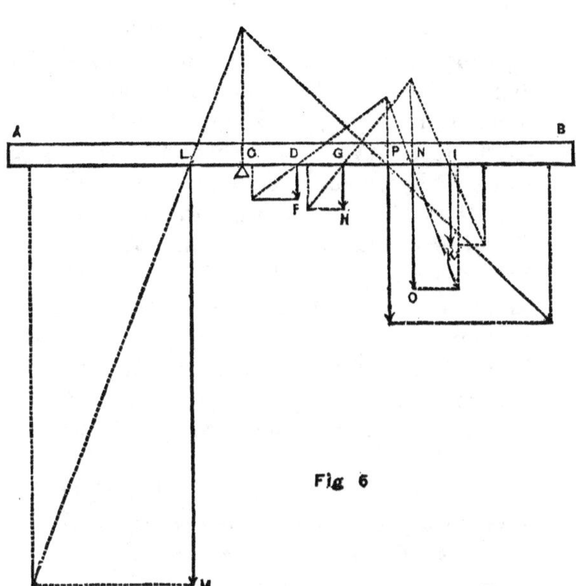

Fig 6

right of C lay off the distance $C\,D$, $C\,G$, $C\,I$, equal to the lever arms of the valve, the lever and the weight respectively; and to the left of C lay off the distance $C\,L$, equal to $C\,D$. Also draw vertical lines, $D\,F$, $G\,H$, $I\,K$, to represent the weights of the valve, lever and ball.

Find by construction a single force, *P R*, which can replace these three forces and produce the same effect. This is done by replacing *G H* and *I K* by a single force, *N O*, and then finding a force, *P R*, which is equivalent to *N O* and *D F*. Then construct the force *L M*, which acts at the point *L* and balances *P R*, or the three forces, *D F*, *G H*, *I K*. *L M* represents the total pressure on the valve in pounds, and this, being divided by the area of the valve in square inches, will give the pressure of the steam, in pounds per square inch, at which the valve will rise.

2*nd*. *To find where to place the weight on the lever of a given safety-valve, in order that the valve may rise with a given steam pressure.*—In Fig. 7, *L M* represents the required pressure in pounds per square inch multiplied by the area of the valve in square inches, *D F* is the weight of the valve, *G H* is the weight of the lever, and *a b* is the weight of the ball, the points of application of all the weights, except that of the ball, being

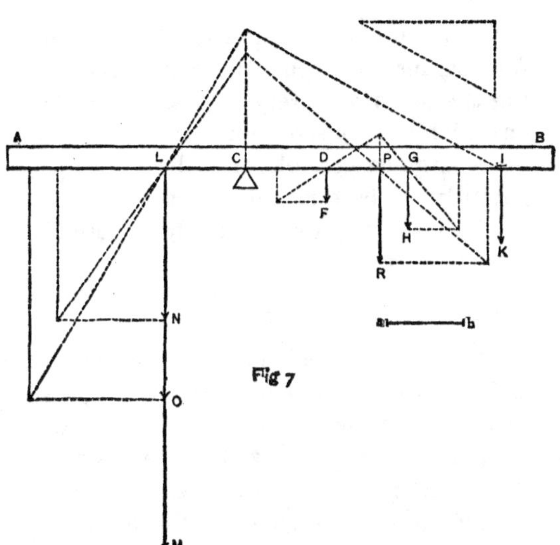

Fig 7

given. Replace *D F* and *G H* by a single force, *P R*, which will produce the same effect, and find *L N*, a force acting at *L*, and balancing *P R*. Subtracting *L N* from the whole force *L M*, there remains *L O*, which must be balanced by the weight of the ball *a b*. Then, by the third construction, previously given, find the point *I* at which the ball must be placed in order to balance *L O*.

In the use of the graphical method for testing the accuracy for proportioning the parts of safety-valves, the scale on which the construction is made should be as large as possible, in order that apparently trivial errors on a drawing may not be magnified greatly in the practical applications.

C.—Analytical Method.

The mode of calculating the parts of safety-valves from given or observed data will now be presented. And first it may be well to say something in regard to the means of ascertaining the weights of the various parts, and the position of the center of gravity of the lever. The most obvious method of determining the weight of any thing is to weigh it, and the center of gravity of a lever is most readily found by balancing a lever upon a knife-edge. It is not always convenient, however, to make these experiments. Sometimes no facilities for weighing the parts are at hand, and in certain cases it may be necessary to make the calculations from such data as can be obtained from a drawing. When the parts can be removed, their weights can be determined approximately by experiment. Immerse each part in water, using a shallow vessel, and mark how much the water rises, which will give, by a simple calculation, the volume of the part in cubic inches. Having ascertained the number of cubic inches in a part, the weight can be obtained by multiplying this number

By 0.2778, if the part is of wrought iron.
By 0.2604, " " " cast iron.
By 0.2917, " " " brass.

The weights so found will generally be only approximate, but will be tolerably accurate, if the experiment is carefully performed.

When only the dimensions of the parts are given, their volumes must be determined by calculation. The valve and weight will ordinarily present no difficulties, as they have forms for which rules are given in most works on mensuration, and their centers of gravity usually coincide with their geometrical centers. Having found their volumes in cubic inches, their weights can be obtained by multiplying these volumes by the weight of a cubic inch of the material as given above. The lever, not unfrequently, is quite irregular in form, and the center of gravity is scarcely ever in the middle of the bar. Whatever its shape, however, its volume can be found with great accuracy by the method to be explained. Make a drawing of the lever in plan and elevation, as shown in Fig. 8, *C D* being the plan, and *E F* the elevation, the form of the lever being much more irregular than is usual in practice. To determine the volume of this lever in cubic inches, it will be necessary to find the area of one face, as, for instance, the elevation and the mean width of the other. In finding the mean width, the area of the face represented on the plan will first be calculated, and then divided by the length. To calculate the area of either of these faces, divide it into any even number of parts,

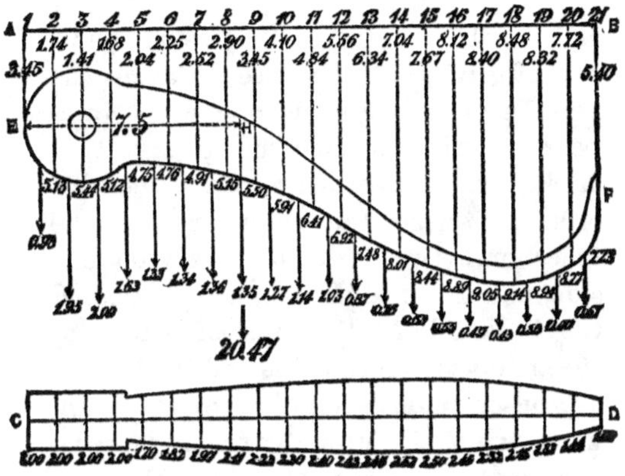

Fig 8

by equi-distant lines drawn perpendicular to any line, assumed as may be most convenient. Thus, in Fig. 8, the elevation is so irregular that a straight line would not lie wholly within it, and *A B* is drawn without, and the elevation is divided into 20 parts, by equi-distant lines perpendicular to *A B*. In the plan, the line *C D* is within the figure, and the same number of equi-distant lines are drawn, perpendicular to *C D*. These perpendicular lines are called ordinates, and it will be observed that as the number of divisions of a face is even, the number of ordinates is uneven. Number the ordinates from first to last, 1, 2, 3, 4, etc., and measure their lengths, either between the boundaries of the faces or between the boundary on one side, and the straight line, which is called the base. Then the area of a face, or the area between the base and the boundary on one side (according to the manner in which the ordinates are measured), can be found by the following rule :

Add together the first and last ordinates, four times the sum of the even numbered intermediate ordinates, twice the sum of the odd numbered ordinates, and multiply the sum so obtained by one-third of the distance between any two consecutive ordinates.

In Fig. 8, the lever is supposed to be 20 inches long, and the distance between ordinates is one inch, so that there are 20 divisions, and 21 ordinates. This example is merely for the sake of illustration. Where great accuracy is required, the ordinates should be taken much closer together. The lengths of the ordinates are shown on the drawing, the dimensions being in inches. The diameter of the hole in the lever is 9-10 of an inch. To find the area of the elevation, the area included between the base *A B* and the boundary of the figure nearest to it will be calculated and subtracted from the whole area between the base and the outside boundary. The following are the calculations :

AREA BETWEEN BASE AND NEAREST BOUNDARY.

Extreme ordinates.	Even-numbered ordinates.	Odd-numbered ordinates.
1 ..... 3.45	2........1.74	3........1.41
21 ..... 5.40	4........1.68	5........2.04
——	6....... 2.25	7........2.52
8.85	8........2.90	9... ....3.45
	10........4.10	11........4.84
	12........5.56	13........6.34
	14........7.04	15........7.67
	16.... ...8.12	17........8.40
	18... ....8.48	19........8.32
	20........7.72	——
	——	44.99
	49.59	

Sum of even-numbered ordinates.......	49.59
Multiply by......................	4
	198.36
Sum of odd-numbered ordinates........	44.99
Multiply by.................. ...	2
	89.98
Add .. .. {	198.36
	8.85
	297.19
Multiply by......................	⅓
Area between base and nearest boundary (square inches)....................	99.06

AREA BETWEEN BASE AND OUTSIDE BOUNDARY.

Extreme ordinates.	Even-numbered ordinates.	Odd-numbered ordinates.
1 ......3.45	2........5.13	3........5.44
21 ......7.23	4........5.12	5 ....4.75
	6........4.76	7........4.91
10.68	8........5.15	9........5.50
	10..... ..5.91	11........6.41
	12........6.92	13........7.48
	14........8.01	15........8.44
	16........8.89	17........9.05
	18........9.14	19........8.94
	20........8.77	60.92
	67.80	

67.80	60.92
4	2
271.20	121.84
	271.20
	10.68
	403.72
	⅓

Area between base and outer boundary (square inches)....................134.57
Subtract.......................... 99.06

Area of elevation of lever, square inches 35.51

AREA OF PLAN OF LEVER.

1.......2.00	2.... ...2.00	3........2.00
21.......1.00	4........2.00	5........1.70
——	6........1.82	7........1.97
3.00	8........2.11	9........2.23
	10........2.30	11........2.40
	12........2.43	13... ....2.45
	14........2.52	15........2.50
	16........2.45	17........2.32
	18... ....2.15	19........1.83
	20........1.44	——
	——	19.40
	21.22	2
	4	——
	——	38.80
	84.88	84.88

	3.00
	126.68
	$\frac{1}{3}$
Area of plan of lever, in square inches..	42.23
Divide by length of lever..............	20)42.23
Average width of plan.....	2.11
Multiply by area of elevation........	35.51
Volume of lever in cubic inches......	74.9261
Deduct volume of hole..............	1.2723
	73.6538
Multiply by weight of a cubic inch of iron..........................	0.2778
Weight of lever in pounds...........	20.47

This method of finding the area of an irregular figure is very useful in many cases that arise in practice. It will be a good plan for the reader to calculate the weight of a lever in the manner here explained, and compare it with the weight as ascertained by trial.

The position of the center of gravity of any lever can also be found by calculation, and the manner of doing it will now be explained. It may be well to give, in the first place, a few definitions and illustrations.

The resultant of two or more forces is a single force, whose action is equal to the combined effect of the several forces which are called components.

The moment of a force with reference to any point is the product arising from multiplying the intensity of the force by the distance of the point from the force measured in a straight line drawn from the point perpendicular to the direction of the force. This distance is called the arm of the force.

When any number of forces act in the same plane, the moment of the resultant is equal to the sum of the moments of those forces which tend to turn a body in the same direction that the resultant does, diminished by the sum of the moments of those forces that tend to turn a body in the opposite direction. This is generally more concisely expressed by saying that the moment of the resultant is equal to the algebraic sum of the moments of the components. In Fig. 9 an illustra-

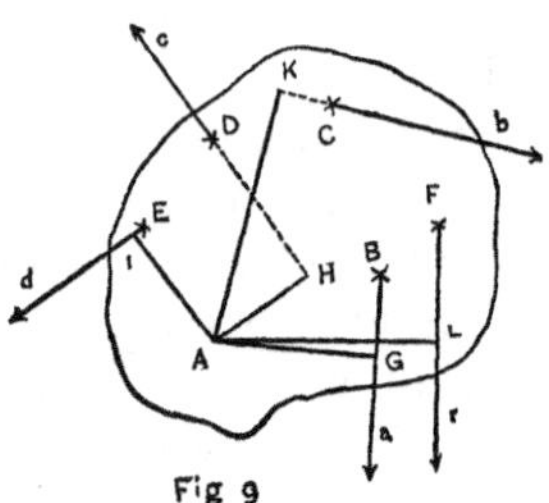

Fig 9

tion of this principle is given. Four forces in the same plane, represented by the darts *a*, *b*, *c*, *d*, are acting at the points *B*, *C*, *D*, *E* of a body which is secured so that it can only turn around the point *A*, and *r* is a single force which will produce the same effect as these four forces, or it is the resultant of *a*, *b*, *c*, *d*. The perpendicular lines drawn from the point *A* to the directions of the several forces represent the arms of these forces. An inspection of the figure will show that the forces *a*, *b*, tend to turn the body in the same direction that the resultant does, while the forces *c*, *d*, tend to turn it in the opposite direction. The analytical statement of the principle given above will, then, be :

$$r \times AL = a \times AG \times b \times AK$$
$$-(c \times AH + d \times AI)$$

From this equation it appears that if all but one of the parts in any example of this kind are known, the other can readily be ascertained. Suppose, for instance, that the intensities of all the forces are known, and all the arms, except that of the resultant, as follows :

$a = 108$	pounds	$AG = 75$	inches.
$b = 114$	"	$AK = 170$	"
$c = 72$	"	$AH = 65$	"
$d = 81$	"	$AI = 75$	"
$r = 150$	"		

From these values the following equation can be formed :

$$150 \times AL = 8{,}100 + 19{,}380 - (4{,}680 + 6075)$$
$$= 27{,}480 - 10{,}755$$
$$= 16{,}725$$
$$\text{and } AL = 16{,}725 \div 150$$
$$= 111\tfrac{1}{2} \text{ inches.}$$

The lever of a safety-valve is acted upon at every point by a force which is the weight of the particle at that point, and the resultant of all these forces is their sum, or the weight of the whole lever, and the point at which it acts is the center of gravity of the lever. This point, then, can be found with sufficient accuracy, by supposing that the lever is divided up into such small parts that the center of gravity of each can be estimated at its geometrical center, without serious error, and its weight calculated as if its faces were trapezoids—so that an equation can be formed with the movements of these forces. In Fig. 8 the lever is divided into 20 parts, and the weight of each is calculated, being shown by the darts on the elevation.

The moments of the force are referred to the point *E*, and the calculation by which the center of gravity is found, is given below :

MOMENT OF WEIGHT OF PART.

1— 2,—0.98 × 0.5 = 0.49
2— 3,—1.95 × 1.5 = 2.93
3— 4,—2.09 × 2.5 = 5.22
4— 5,—1.63 × 3.5 = 5.71
5— 6,—1.33 × 4.5 = 5.98
6— 7,—1.34 × 5.5 = 7.37
7— 8,—1.36 × 6.5 = 8.84
8— 9,—1.35 × 7.5 = 10.13

9—10,—1.27	× 8.5 =	10.79
10—11,—1.14	× 9.5 =	10.83
11—12,—1.03	× 10.5 =	10.82
12—13,—0.87	× 11.5 =	10.00
13—14,—0.76	× 12.5 =	9.50
14—15,—0.63	× 13.5 =	8.51
15—16,—0.55	× 14.5 =	7.97
16—17,—0.49	× 15.5 =	7.60
17—18,—0.43	× 16.5 =	7.09
18—19,—0.36	× 17.5 =	6.30
19—20,—0.40	× 18.5 =	7.40
20—21,—0.51	× 19.5 =	9.95
		153.43

Calling EH the distance from the end of the lever to the center of gravity, and recollecting that the resultant is equal to the weight of the lever,

$$20.47 \times EH = 153.43:$$
$$\text{Whence, } EH = 7\tfrac{1}{2} \text{ inches.}$$

This mode of determining the center of gravity of a lever is quite tedious, if the parts are taken small enough to secure great accuracy, but it is useful in case the plan by trial is impossible.

The condition that a safety-valve shall

rise at a given pressure of steam requires, as has already been stated, that the forces tending to hold the valve down shall be just balanced by the tendency of the opposing forces to raise it. In Fig. 1 there are three forces holding the valve down, the weight of the ball, *W*, the weight of the lever, *L*, and the weight of the valve *V*, and these must be balanced by the pressure of the steam, *P*. When any number of forces balance each other, the sum of the moments of the forces which act in one direction is equal to the sum of the moments of the forces which act in the other direction. Hence the following equation will represent the action of the forces when the valve is just balanced :

$$P \times p = W \times w + L \times l + V \times p.$$

If any seven of the quantities in this equation are known, the other can be determined. The equation can, however, be changed a little, so as to be more generally useful :

Let S= pressure of steam in pounds per

square inch, D = diameter of valve in inches ; then,

$$P = 0.7854 \times D^2 \times S,$$

and the equation becomes :

$$0.7854 \times D^2 \times S \times p$$
$$= W \times w + L \times l + V \times p.$$

For the ordinary cases that arise in practice the equation will take the following forms ;

1st. *To find the pressure per square inch at which a given valve will open* :

$$S = \frac{W \times w + L \times C + V \times p}{0.7854 \times D^2 \times p}$$

2d. *To find where to place the weight on a given safety valve, so that it shall open at a given pressure per square inch* :

$$W = \frac{0.7854 \times D^2 \times S \times p - L \times l - V \times p}{W}$$

3d. *To find what diameter a safety valve must have, all the other parts being known, to open at a given steam pressure :*

$$D=\sqrt{\left(\frac{W\times w+L\times l+V\times p}{0.7854\times S\times p}\right)}$$

As many who are familiar with arithmetic, but do not understand algebraic expressions, may like to use these rules, they are translated below :

1st. *To find the pressure per square inch at which a given valve will open.*

Measure the following distances horizontally from the fulcrum to :

1. The center of the valve stem.
2. The center of the weight.
3. The center of gravity of the lever, or the point on which it will balance if placed upon a knife-edge.

Measure the diameter of the valve, and determine its area, either by a table or by multiplying the square of the diameter by 0.7854.

Find the weight of :

1. The valve.
2. The lever.
3. The ball.

Multiply—

(1). The weight of the ball by its horizontal distance from the fulcrum.

(2). The weight of the lever by its horizontal distance from the fulcrum.

(3). The weight of the valve by its horizontal distance from the fulcrum.

(4). The area of the valve by its horizontal distance from the fulcrum.

Add together the first three products and divide the sum by the fourth product.

Example.—A given safety-valve has a weight of 50 pounds 24 inches from the fulcrum, the lever weighs 6 pounds, and its center of gravity is 15 inches from the fulcrum; the weight of the valve is 2 pounds, and its center is 4 inches from the fulcrum. The diameter of the valve is 2 inches. At what pressure will the valve begin to rise?

Square of diameter................ 4
Multiply by... ....0.7854

Area of valve in square inches ...3.1416
(1). 50 times 24 is 1,200
(2). 6 " 15 " 90
(3). 2 " 4 ".... . 8
(4.) 3.1416 " 4 " 12.5664

The sum of (1), (2) and (3) is 1,298,

which divided by 12.5664 (the fourth product) is 103.03, the pressure in pounds per square inch at which the valve will open.

2. *To find where to place the weight on a safety-valve, so that it shall open at a given pressure of steam:*

Multiply—

(1). The weight of the lever by the horizontal distance of its center of gravity from the fulcrum.

(2). The weight of the valve by its horizontal distance from the fulcrum.

(3). The area of the valve by the pressure of steam in pounds per square inch, and by the horizontal distance of the valve from the fulcrum.

Add together the first two products, subtract their sum from the third product, and divide the difference by the weight of the ball.

Example.—The ball of a safety-valve weighs 100 pounds, the lever weighs 10 pounds, the valve weighs 2 pounds and has a diameter of 3 inches. The distance of the center of gravity of the lever from the fulcrum is 25 inches, and the distance of the center of the valve from the fulcrum is 5 inches. How far from the fulcrum must the valve be placed, in order that the lever may open at a pressure of 100 pounds?

Area of valve........ 7.07 square inches.
(1). 10 times 25 is........ 250
(2). 2 " 5 ". 10
(3). 7.07 " 100 " 707
707 " 5 ".. 3,535

Adding together products (1) and (2) we have as their sum 260; subtracting this from 3,535, the third product, we have 3,275. Dividing this difference by 100, the weight of the ball, we have 32.75, or 32¾ inches as the distance from fulcrum to ball.

3d. *To find what diameter a safety-valve must have, the other parts being known, to open at a given steam pressure:*

Multiply,

(1). The weight of the ball by its horizontal distance from the fulcrum.

(2). The weight of the lever by the

horizontal distance of its center of gravity from the fulcrum.

(3). The weight of the valve by the horizontal distance of its center from the fulcrum.

(4). The pressure of steam in pounds per square inch by the horizontal distance of the valve from the fulcrum, and by the number 0.7854.

Add together the first three products, divide their sum by the fourth product, and take the square root of the quotient.

Example.—Weight of ball, 60 pounds; lever, 7 pounds; valve, 3 pounds. Distances from fulcrum: ball, 30 inches; center of gravity of lever, 16 inches; center of valve, 3 inches. Pressure of steam, 70 pounds per square inch. What should be the diameter of the valve?

(1).	60	times	30	is.....	1,800
(2).	7	"	16	".....	112
(3).	3	"	3	".....	9
(4).	70	"	3	".....	210
	210	"	0.7854	"....	164.934

The sum of the first three products (1,800, 112 and 9) is 1,921. Dividing this sum by 164.934 (the fourth product), we have 11.647 inches. The square root of this number is 3.41 + inches, which by the rule is the required diameter of the valve.

Instead of weights, springs are sometimes employed to hold down safety-valves. The calculations involving merely the condition that the valve shall open at a given pressure, are the same as those previously employed, except that the tension of the spring is to be substituted for the weight of the ball, this tension being first determined by experiment. Having ascertained the dimensions of a spring most suitable for a given case, it is easy to calculate the dimensions of a spring for any other case. Knowing, for instance, the cross-section of a spring adapted to one pressure on the valve, make a proportion, thus :

Cross section of given spring.	:	Cross section of required spring.	::	Pressure on first valve.	:	Pressure on second valve.

The pitch of the second spring, or the

distance from the center of one coil to the center of the next, is found by the following proportion :

Pitch of given spring : Pitch of required spring. : : { Side of a square equal in area to section of first spring } : { Side of a square equal in area to section of 2d spring }

Example.—It is a common practice in proportioning the parts of direct-loaded spring safety-valves to use a spring of the following dimensions : for a valve 3 inches in diameter and 100 pounds steam pressure, a spring made of square steel, having a cross section of ¼ of a square inch, and a pitch of one inch, the side of a square equal in area to the cross section being, of course, ½ inch. What should be the proportions of a similar spring for a valve 5 inches in diameter, and a steam pressure of 50 pounds per square inch?

Area of first valve...............	7.07 in.
Multiply by steam pressure......	100
Total pressure on first valve...	707

Area of second valve............19.64 in.
Multiply by steam pressure...... 50

Total pressure on second valve. 982

To find cross section of required spring :

$$0.25 : \frac{\text{Cross section}}{\text{of second spring}} :: 707 : 982$$

and by the rule for proportion, the product of the two extremes (245.5), divided by the third term (707), will give the second term. Now 245.5 divided by 707 is 0.34724 in., equals cross section of second spring. The square root of 0.32724 is 0.589+, and this is the side of a square equal in area to cross section of second spring.

To find pitch of second spring :

$$1 : \frac{\text{Pitch of}}{\text{second spring}} :: 0.5 : 0.589.$$

Here we solve the problem by multiplying 1 by 0.589 and dividing the product by 0.5, which gives 1.178 inches for the pitch of second spring.

III. Proper Diameter for a Safety-Valve.

There are seven rules, commonly quoted by different authorities, for determining the area of a safety-valve. They are given below, with an example to illustrate their use:

Let

G=square feet of grate surface in boiler.

H=square feet of heating surface in boiler.

C=pounds of coal burned per hour.

W= " " water evaporated per hour.

P=pressure of steam, as shown by gauge.

A=area of safety-valve in square inches.

When the area is known, the diameter of the valve can be found by dividing the area by the number 0.7854, and extracting the square root of the quotient.

1. *United States Rule.*—Allow one

square inch of area in the valve for 25 square feet of heating surface in the boiler, or

$$A = \frac{H}{25}$$

2. *English Rule.*—For boilers with natural draft, allow half a square inch of area in the valve for each square foot of grate surface, or

$$A = \frac{G}{2}$$

3. *French Rule.*—1. Multiply the grate surface by the number 22.5.

2. Add the number 8.62 to the steam pressure.

3. Divide the first quantity by the second. The quotient will be the area of valve, or

$$A = \frac{G \times 22.5}{P + 8.62}$$

4. *Molesworth's Rule.*—Allow a valve area of eight-tenths of an inch for each square foot of grate surface, or

$$A = G \times 0.8.$$

5. *Professor Thurston's First Rule.*—1. Multiply the pounds of coal burned per hour by the number 4.

2. Add the number 10 to the steam pressure.

Divide the first quantity by the second, or

$$A = \frac{4\ C}{P+10}$$

6. *Professor Thurston's Second Rule.*—1. Multiply the heating surface by the number 5.

2. Add the number 10 to the steam pressure, and multiply the sum by the number 2.

Divide the first quantity by the second, or,

$$A = \frac{5\ H}{2(P+20)}$$

7. *Professor Rankine's Rule.*—Allow a valve area of six-thousandths of an inch for each pound of water evaporated per hour, or,

$$A = 0.006\ W$$

In many cases these rules give widely different results, as will be illustrated by their application to an example.

It is required to find the area of a safety-valve for a boiler having 15 square feet of grate surface, 472.5 square feet of heating surface, carrying 40 pounds of steam, burning 210 pounds of coal, and evaporating 1,470 pounds of water per hour.

1. *United States Rule:*

472.5÷25=18.9 square inches=area required.

2. *English Rule:*

15÷2=7.5 square inches=area required.

3. *French Rule:*

40+8.62=48.62.

15×22.5=337.5.

337.5÷48.62=6.94 square inches=area required

4. *Molesworth's Rule:*

15×0.8=12 square inches=area required.

5. *Professor Thurston's First Rule:*

40+10=50.

210×4=840.

840÷50=16.8 square inches=area required.

6. *Professor Thurston's Second Rule:*

40+10=50.

50 × 2 = 100.

472.5 × 5 = 2,362.5.

2,362.5 + 100 = 23.63 sq. inches = area required.

7. *Professor Rankine's Rule:*

1,470 × 0.006 in. = 8.82 sq. inches = area required

It is not remarkable that these rules should give such varying results, when it is remembered that the performance of different boilers of precisely the same dimensions varies greatly. A safety-valve should have such proportions that it can permit all the steam to escape that a boiler can generate when forced to the utmost extent. Hence its area depends upon:

1. *The amount of steam to be discharged in a given time.*

2. *The lift of the valve.*

3. *The velocity with which the steam escapes.*

There is such a difference in the amount of coal burned per square foot of grate per hour in different boilers, and the amount of water evaporated per pound of coal, that it would seem impossible to use the same constants in a formula for all cases. It is believed, however, that the following estimates give a most liberal allowance on the side of safety :

	Pounds of coal burned per square foot of grate per hour.	Pounds of water evaporated per pound of coal.
Stationary and marine boilers with natural draft....	15	9
Stationary and marine boilers with forced draft....	30	7
Locomotive boilers..........	100	6

On these assumptions, the pounds of water evaporated, and consequently the weight of steam that the safety-valve would have to release per hour for each square foot of grate surface would be:

For stationary and marine boilers with natural draft..........................	135
For stationary and marine boilers with forced draft..........................	210
For locomotive boilers...........	600

These figures, as has been stated, represent much better than average performance, and can readily be modified to suit any particular case. Of course, in constructing a formula for general use, in the absence of precise data, and in view of the varying results obtained from boilers of the same kind, the estimate should be taken high enough to include all cases.

One element of the proposed formula for the diameter of a safety-valve for any given boiler, then, will be the number of square feet of grate surface in the boiler, multiplied by a constant.

The amount of opening afforded by a safety-valve for the escape of steam depends not only on its diameter, but also on the distance that the valve lifts. In order that the area for escape shall be equal to the area of the valve, it is necessary that the lift should be one-quarter of the diameter of the valve.

A valve without bevel, Fig. 10, evi-

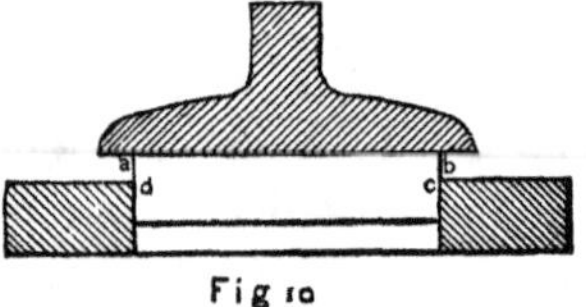

Fig 10

dently gives an opening, when raised, equal to the surface of a cylinder having the same diameter, *a b*, as that of the valve, and a height equal to the lift, *c b*. *Hence the rule for finding the opening of a flat valve, due to a given lift, will be as follows:* multiply the diameter of the valve by the lift, and by the number 3.1416.

Example.—A flat valve 3 inches in diameter lifts 1-16 of an inch. What is the area of the opening?

Lift..................................	0.0625
Multiply by..........................	3.1416
	0.196+
Multiply by..........................	3
Area of opening in square inches....	0.589+

If the valve has a bevel, as shown in Fig. 11, until it lifts clear of the seat the opening will be equal to the surface of the frustum of a cone, of which the diameter, *a b*, of the upper base is the same as that of the valve, the slant height, *b e*, is the perpendicular distance between the lower edge of the valve and

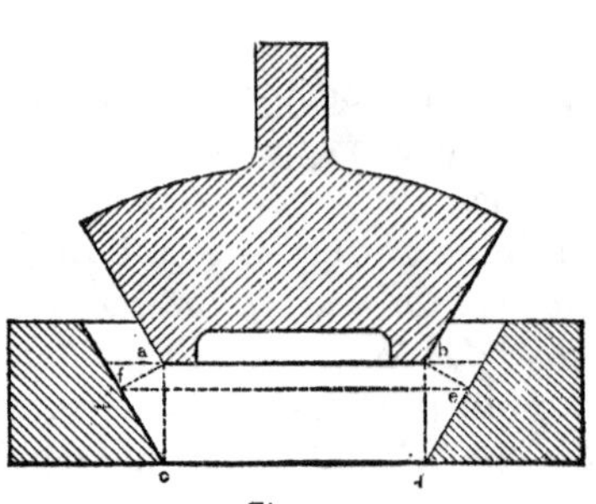

Fig 11

the seat, and the diameter of the lower base, *f e*, is the diameter of the seat, measured at the intersection of perpendiculars *b e*, *a f*, from opposite points of the lower edge of the valve to the seat. The bevel or inclination of the valve is the angle of inclination to a vertical line, *f c a*, or *e d b*.

To find the amount of opening afforded by a valve with beveled seat for any lift less than the depth of the seat.

(1). Multiply the diameter of the valve by the lift, by the sine of the angle of inclination, and by the number 3.1416.

(2). Multiply the square of the lift by the square of the sine of the angle of inclination, by the cosine of this angle, and by the number 3.1416.

Add these two products.

Example.—The diameter of a safety-valve is $2\frac{1}{2}$ inches, the seat is $\frac{5}{8}$ of an inch deep, and has a bevel of 25 degrees. What is the area of opening, for a lift of $\frac{1}{4}$ of an inch?

Sine of 25°........................	0.423
Multiply by lift......	0.25
	0.106—
Multiply by..........................	3.1416
	0.333
Multiply by diameter of valve........	2.5
1st Product.....................	0.83+
Square of sine of 25°......	0.179
Multiply by cosine of 25°.............	0.906—
	0.162+
Multiply by square of lift............	0.0625
	0.0101+
Multiply by............................	3.1416
2d product..........................	0.03+
Add 1st product............... ...	0.83+
Area of opening in square inches.....	0.86+

It often happens that the depth of the bevel is very slight, so that the valve lifts clear of the seat. In such a case the opening must be computed for a beveled valve with a lift equal to the depth of the seat, and for a flat valve with the remainder of the lift.

Example.—The diameter of a valve is 4 inches, the bevel is 35°, and the depth of seat ¼ of an inch. What is the area of opening for a lift of ⅝ of an inch?

First calculate the the area of opening for a beveled valve, with a lift of ¼ inch, the depth of the seat.

Sine of 35°	0.5745
Multiply by	3.1416
	1.8
Multiply by lift	0.25
	0.45
Multiply by diameter of valve	4
First product	1.80
Square of sine of 35°	0.329
Multiply by cosine of 35°	0.819
	0.269
Multiply by	3.1416
	0.85

Multiply by square of lift............. 0.0625

Second product... .... 0.05
Add first product..................... 1.80

Area of opening, in square inches, in lifting clear of seat........... 1.85

Next calculate the amount of opening due to the lift above the seat, which in this case is $\frac{1}{8}$ of an inch:

Lift.................. .... 0.125
Multiply by........................... 3.1416

0.3927
Multiply by diameter of valve... 4

Area of opening due to lift above seat, in square inches 1.57
Add opening due to lifting clear of seat 1.85

Total area of opening, in square inches.......................... 3.42

The most usual bevel for valves is an angle of 45°, and 30° is also frequently adopted. As a matter of convenience, the rules have been somewhat simplified for these cases.

To find area of opening for a lift less than depth of seat, when the bevel of the valve is 45 degrees.

1. Multiply the diameter of the valve by the lift, and by the number 2.22.

2. Multiply the square of the lift by the number 1.11.

Add these two products.

Example.—What is the area of opening of a 2-inch valve, for ½-inch lift; depth of seat, ¼ inch; bevel of valve 45 degrees?

1. Opening in lifting clear of seat, ¼ inch=0.25 inch:

(1). 2×2.22×¼=1.11.
(2). 0.25×0.25×1.11=0.07.

Adding the products (1) and (2)

1.11+0.07=1.18 square inches in lifting clear of seat.

2. Opening for lifting above seat, ¼ inch:

3.1416×¼×2=1.57 square inches, in lifting above seat.
1.57+1.18=2.75 square inches=total area of opening.

To find area of opening for a lift less than depth of seat, when the bevel of the valve is 30°.

1. Multiply the diameter of the valve by the lift, and by the number 1.57.

2. Multiply the square of the lift by the number 0.68.

3. Add the two products.

Example.—A 4-inch valve has a bevel of 30°, and the depth of seat is ¼ inch. What is the area of opening for a lift of 3-16=0.1875 of an inch?

(1). $4 \times 0.1875 \times 1.57 = 1.18$

(2.) $0.1875 \times 0.1875 \times 0.68 = 0.035 \times 0.68 = 0.02$

(3). $1.18 + 0.02 = 1.2$ square inches= area of opening.

The analytical expressions for the preceding rules are given below, together with the values of the sines and cosines of angles from 20° to 50°.

Let
D=diameter of valve in inches.
l=lift " "
a=angle of bevel.
d=depth of seat, in inches
A=area of opening, in square inches.

1. *When there is no bevel:*

$$A=3.1416\times l\times D$$

2. *Beveled valve, lift less than depth of seat.*

$$A=3.1416\times[D\times l\times\sin a+l^2\times(\sin a)^2\times\cos a]$$

3. *Beveled valve, lift greater than depth of seat.*

$$A=3.1416\times[D\times d\times\sin a+d^2\times(\sin a)^2\times\cos a+D\times(l-d)]$$

4. *Bevel of* 45°, *lift less than depth of seat :*

$$A=2.22\times D\times l+1.11\times l^2$$

5. *Bevel of* 45°, *lift greater than depth of seat.*

$$A=2.22\times D\times d+1.11\times d^2+3.1416\times D\times(l-d)$$

6. *Bevel of* 30°, *lift less than depth of seat :*

$$A=1.57\times D\times l+0.68\times l^2$$

7. *Bevel of* 30°, *lift greater than depth of seat :*

$$A=1.57\times D\times d+0.68\times d^2+3.1416\times D\times(l-d)$$

Angle.	Sine.	Cosine.	Angle.	Sine.	Cosine.
20°	.342	.940	36°	.588	.809
21°	.358	.934	37°	.602	.799
22°	.375	.927	38°	.616	.788
23°	.391	.921	39°	.629	.777
24°	.407	.914	40°	.643	.766
25°	.423	.906	41°	.656	.755
26°	.438	.899	42°	.669	.743
27°	.454	.891	43°	.682	.731
28°	.469	.883	44°	.695	.719
29°	.485	.875	45°	.707	.707
30°	.500	.866	46°	.719	.695
31°	.515	.857	47°	.731	.682
32°	.530	.848	48°	.743	.669
33°	.545	.839	49°	.755	.656
34°	.559	.829	50°	.766	.643
35°	.574	.819			

It will now be necessary to find the velocity with which steam escapes through the opening afforded by a valve. Professor Rankine determined that the number of pounds of steam that will flow through an orifice having an area of one square inch, into the atmosphere, in a second, is one-seventieth of the absolute pressure of the steam; and this result has been confirmed by the experiments of Mr. Napier. It should be remarked that this rule is only approximate, as the constants vary slightly for different pressures of steam, but it is sufficiently accurate for practical purposes. The absolute pressure of steam is the pressure as shown by a steam gauge, increased by the pressure of the atmosphere, which is, on an average, 14.7 pounds per square inch. *From these principles, a rule can be constructed for the area of opening required in any given case.*

Multiply the absolute pressure of steam in the boiler by the number 51.43, and divide the number of pounds of water evaporated per hour by this product.

Example.—Suppose that a boiler evaporates 2,000 pounds of water per hour, what should be the area of opening afforded by the safety-valve, so that the pressure shall not exceed 100 pounds per square inch?

Pressure by gauge..................	100.0 lbs.
Add..........................	14.7
Absolute pressure.............	114.7 lbs.
Multiply by.....................	51.43
	5,899+

2,000÷5,899=0.339+ =area of opening in square inches. Knowing the area of opening required, if the lift is fixed, the diameter of the valve that will furnish the requisite opening with the given lift can be calculated. There are three general cases for which rules will be given :

1st. *When the valve has no bevel*—Multiply the lift of the valve by the number 3.1416, and divide the area of opening by this product.

Example—A safety valve is required to have one inch area of opening, with a lift of ¼ inch. What should be its diameter?

Lift..................................	0.25 in.
Multiply by..........................	3.1416
	0.7854
0.7854) 1.0000	
Diameter of valve.......... 1.27+inches.	

2d. *When the valve is beveled, and the lift is less than the depth of seat:*

1. Multiply the square of the lift by the square of the sine of the angle of inclination, by the cosine of this angle, and by the number 3.1416.

2. Multiply the lift by the sine of the angle of inclination, and by the number 3.1416.

3. Subtract the first product from the area of opening, and divide the difference by the second product.

Example—A safety valve has a bevel of 32°, and the depth of seat is ½ inch. It is to lift ⅜ of an inch, and give an opening of 1½ square inches. What should be its diameter !

Square of lift..........................	0.141
Square of sine of 32°......	0.281
Their product..........................	0.04
Cosine of 32°..........................	0.848
Product of last two....................	0.034
Multiply by..............	3.1416
And we have...........................	0.107
Area of opening......................	1.500
Subtract.............................	0.107
	1.393
Lift................................	0.375
Sine of 32°................	0.53
Their product...	0.199

Multiply by..................... .. 3.1416

And we have......................... 0.625

Dividing 1.393 by 0.625 we have

Diameter of valve=2.23+inches.

3d. *When the valve is beveled, and the lift is greater than the depth of seat:*

1. Multiply the square of the depth of seat by the square of the sine of the angle of inclination, by the cosine of the angle, and by the number 3.1416.

2. Multiply the depth of seat by the sine of the angle of inclination, and by the number 3.1416.

3. Multiply the difference between the lift and depth of seat by the number 3.1416.

4. Subtract the first product from the area of opening, and divide the difference by the sum of the second and third products.

Example—A safety valve has a bevel of 33°, a depth of seat of $\frac{1}{4}$ inch, and is required to give an area of opening of 2 inches, with a lift of $\frac{1}{2}$ inch. What should be its diameter?

Square of depth of seat	0.0625
Square of sine of 33°	0.297
Product	0.019
Cosine of 33°	0.839
Product	0.016
Multiply by	3.1416
Product	0.05
Area of opening	2.00 in
Subtract	0.05
Difference is	1.95
Depth of seat	0.25
Sine of 33°	0.545
Their product	0.136
Multiply by	3.1416
And we have	0.427
Lift	0.50
Subtract depth of seat	0.25
The difference is	0.25
Multiply by	3.1416
And the product is	0.785
Add	0.427
	1.212

Now 1.95 divided by 1.212 gives the diameter of valve=1.61 inches, nearly.

It may be useful to give a summary of the foregoing rules.

To find the proper diameter for a safety-valve:

1. The number of pounds of water evaporated per hour is equal to the area of the grate in square feet multiplied by,

135.	For stationary and marine boilers with natural draft;
210.	For stationary and marine boilers with forced draft;
600.	For locomotive boilers.

2. Find the area of opening required to discharge this weight of steam per hour, the pressure of steam being given.

3. Find the diameter of a valve that will afford the required opening, with a given lift.

Example.—A stationary boiler with natural draft has 36 square feet of grate surface, and the pressure of steam is 90 pounds. What should be the diameter of the safety-valve, the valve to have a bevel of 45°. a depth of seat of $\frac{3}{8}$ of an inch, and to lift $\frac{1}{4}$ inch?

Grate surface....................	3.6
Multiply by........................	135
Pounds of water evaporated per hour.	4,860
Steam pressure, per gauge..........	90
Add...............................	14.7
Total pressure..................	104.7
Multiply by.......................	51.43
	5,384.7

4,860÷5,384.7

Gives area of opening in square inches=	0.903
Square of lift.................. ...	0.0625
Square of sine of 45°............	0.5
Their product......................	0.031
Multiply by cosine of 45°...........	0.707
And we have.................... ...	0.022
Which, multiplied by...............	3.1416
Gives.............................	0.069
Area of opening....	0.903
Subtract...........................	0.069
The difference is...................	0.83
Lift.................	0.25
Multiply by sine of 45°.............	0.707
Product......	0.177
Multiply by........................	3.1416
Product............................	0.556

$0.834 \div 0.556 = 1.5$ inches = diameter of valve.

The algebraic expressions for the preceding rules are as follows :

Let

D = diameter of valve, in inches.

l = lift " "

a = angle of bevel.

d = depth of seat, in inches.

A = area of opening in square inches.

W = pounds of steam to be discharged through the safety-valve per hour.

P = pressure of steam, in pounds per square inch, as shown by gauge.

p = absolute pressure of steam, in pounds per square inch.

$$p = P + 14.7$$

$$A = \frac{W}{51.43 \times p}$$

1. *When the valve has no bevel:*

$$D = \frac{A}{3.1416 \times l}$$

2. *When the valve is beveled, and the lift is less than the depth of seat:*

$$D=\frac{A-3.1416\times l^2\times(\sin a)^2\times\cos a}{3.1416\times l\times\sin a}$$

3. *When the valve is beveled, and the lift is greater than the depth of seat:*

$$D=\frac{A-3.1416\times d^2\times(\sin a)^2\times\cos a}{3.1416\times(d\times\sin a+l-d)}$$

4. *When the bevel of the valve is* 45°, *and the lift is less than the depth of seat:*

$$D=\frac{A-1.11\times l}{2.22\times l}$$

5. *When the bevel of the valve is* 45°, *and the lift is greater than the depth of seat:*

$$D=\frac{A-1.11\times d^2}{2.22\times d+3.1416\times(l-d)}$$

6. *When the bevel of the valve is* 30°, *and the lift is less than the depth of seat:*

$$D=\frac{A-0.68\times l^2}{1.57\times l}$$

7. *When the bevel of the valve is* 30°,

and the lift is greater than the depth of seat:

$$D=\frac{A-0.68\times d^2}{1.57\times d+3.1416\times(l-d)}$$

It would be a good plan for those who have charge of steam boilers to experiment with the safety-valves, and see whether they conform to these rules. It will be easy to measure the lift of a valve, when it opens under the highest pressure of steam that is to be carried on the boiler to which it is attached, and the other data needed can be readily ascertained. To determine the bevel of a valve, measure its greatest and least diameters, together with the depth of seat. Then divide half the difference between the diameters by the depth of seat, and the quotient is the tangent of the angle of inclination. The algebraic expression is as follows :

Let
T = greatest diameter of valve, in inches.
t = least " " "
d = depth " " "

a = angle of inclination.

$$\tan a = \frac{T - t}{2\,d}$$

The tangents of angles from 20° to 50° are given in the accompanying table, and it will be sufficiently accurate in any example to take the angle corresponding to the number in the table that is nearest to the calculated quotient.

Example.—The greatest and least diameter of a valve are 4 6-10 and 4 inches, respectively, and the depth is ½ inch. What is the bevel?

Greatest diameter....................	4.6
Least "	4.
	2)0.6
	0.5)0.3
Tangent of angle of inclination......	0.6

From the table it appears that the angle corresponding to this is nearly 31°.

Table of Tangents from 20 to 50 Degrees.

Angle.	Tangent.	Angle.	Tangent.
20°	.364	36°	.727
21°	.384	37°	.754
22°	.404	38°	.781
23°	424	39°	.810
24°	.445	40°	.839
25°	.466	41°	.869
26°	.488	42°	.900
27°	.510	43°	.933
28°	.532	44°	.966
29°	.554	45°	1.000
30°	.577	46°	1.036
31°	.601	47°	1.072
32°	.625	48°	1.111
33°	.649	49°	1.150
34°	.675	50°	1.192
35°	.700		

IV. Proper Form for a Safety-Valve in order that it shall have a Given Lift.

When a safety-valve is raised by the pressure of the steam in a boiler it usually exposes a greater area to the pressure. Hence, if the valve were loaded with a weight it might be expected that it would open wide at once from the unbalanced pressure due to the action of the steam on the increased area. It is found in practice, however, that the ordinary safety-valve only rises very slightly when the pressure for which it is set is reached, and that it is necessary for this pressure to be increased considerably to raise the valve much higher. In a previous part of this paper it was stated that as soon as a safety-valve was opened a new set of conditions was to be considered. With the opening of the valve, steam begins to escape from the boiler with a high velocity, and in its escape it has to overcome the resistance afforded by the opening. In this way its pressure is reduced, near the orifice, so that, although there is a greater area of valve to be acted on by the steam, the pressure is reduced. How much the pressure will be reduced in any given case can only be determined by experiment, since the data for a theoretical calculation cannot be given. Some experiments made in England showed that in the case of a three-inch valve, relieving a boiler at a pressure of 100 pounds per square inch, the pressure under the valve was 25 pounds per square inch. It is probable that the pressure under a valve, when opened, will vary with the form of valve and the pressure in the boiler, as a change in the valve may vary the resistance opposed to the escape of steam, and a change of steam pressure effects a corresponding change in the velocity with which the steam issues from the orifice.

In the case of a safety-valve kept down by a spring, it must be evident that unless arrangements are made for increasing the upward pressure, when the valve is opened it will lift but slightly, since every increase of lift requires additional force for the extension or compression of the spring. But in whatever manner the valve is loaded, it would seem that the first step in proportioning it would be to find what pressure is acting when the valve is opened. It will then be possible, knowing the resistance that must be overcome to give the desired lift, to afford such an increase of area in the open valve that the steam pressure will balance the resistance. It must be remembered in this connection, that it is desirable to have the valve close promptly as soon as the pressure is reduced, and unless some special provision is made for this, the valve, though it may open wide from the pressure due to the increased area, will be prevented, from the same cause, from closing until the pressure is greatly reduced. As before remarked, experiment is necessary in order to give data by which the desired results can be effected, and experiments in this direction have not been sufficiently extended to enable general rules to be deduced. It may be interesting and instructive, however, to call attention to examples of valves that give good results in practice, and explain, as far as possible, the principles on which they act. It is not considered necessary to illustrate all the safety-valves in the market, but only such prominent ones as embody distinct principles. And first it may be well to give the results of some experiments made by a commission appointed by Congress to examine life-saving inventions, in 1868. This commission received 14 safety-valves for examination, and tested them by attaching them to a steam boiler and observing their action. It is reasonable to suppose that the valves offered for competition represented the most prominent and successful ones in the market; but it will be seen that many of them were far from fulfilling all the conditions required of a good safety-valve. The accompanying table gives the essential features of the test, the different valves being represented by numbers.

Valve.	Locked or Open.	Open at / Set to	Blew off at	Closed at
No. 1.	Open.	107	115	106
	Locked.	107	117	114
No. 2.	Open.	53	59	58
	Locked.	53	60	57
No. 3.	Open.	40	58	52
No. 4.	Open.	45	49	25 or 24, but continued blowing, until shut off by hand.
	Open.	45	46	35
	Locked.	45	45	35

Valve.	Locked or Open.	Open at Set to.	Blew off at	Closed at
No. 5.	Open.	42	50	50
	Open.	82	91	89
	Locked.	82	91	89
No. 6.	Open.	35	45, began to blow a little, but was shut off at 54, not havlng blown freely, even at that point.	
No. 7.	Open.	90	93	74, and did not cease blowing.
	Locked.	90	90	85
	Open.	50	50	45
	Open.	50	50	46
	Locked.	50	50	46
No. 8.	Open.	50	60	50
	Locked.	42	45	38
No. 9.	Open.	103	106	92
	Locked.	103	106	97
No. 10	Open.	75	80	74
	Locked.	75	80	73
No. 11	Open.	100	99	Failed to close, shut off at 85.
	Locked.	100	98	95, leaking badly.

Valve.	Locked or Open.	Open at Set to.	Blew off at	Closed at
No. 12	Open.	130	133	Shut off at 114.
	Locked.	130	136	Shut off at 115.
	Locked.	135	137	121
No. 13	Open.	136	137	138
	Locked.	136	136	137
No. 14	Open.	30	31	30
	Locked.	30	31	30½

In some experiments made by the Academy of Sciences, in France, and the Franklin Institute in this country, a valve of the form shown in Fig. 12 was employed. It will be apparent that as soon as the valve is raised a little, the weights will move on the lever, causing the valve to open wide at once. A valve of this kind, however, is not suitable for general practice, since a sudden opening to the full capacity is apt to empty the boiler of both steam and water, and, moreover, there is no provision made for the automatic closing of the valve by a slight reduction of the pressure.

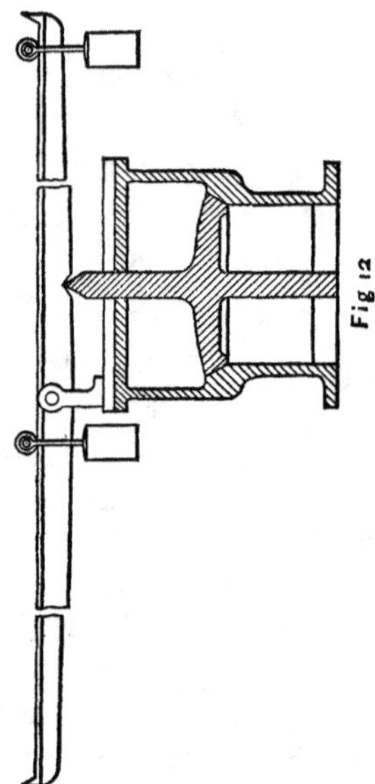

An ordinary valve, if not kept constantly in use frequently sticks fast to the seat, so that the pressure must be raised beyond the point for which it is set before it will open. The friction of the joints on the lever also frequently increases the resistance to opening very considerably. A valve called by the inventor the "positive safety-valve" is sketched in Fig. 13. The seat is spherical, and the weight is within the boiler,

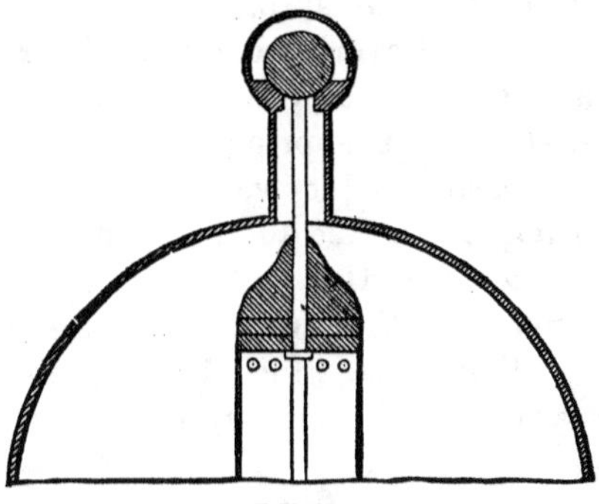

Fig 13

a tube on the end of the valve-stem extending down into the water. There are no joints to cause friction in the valve, and the motion of the water produces a movement in the valve, so that it is continually changing its position on the seat, and is not liable to stick. By a proper proportion of seat it is possible to expose enough additional area of valve, when opened, to give any desired lift.

Instead of pin connections for the levers of safety-valves, it is common to make the levers turn on knife edges, thus reducing the friction. In the case of valves loaded by springs, it has been mentioned that every increase of lift requires an increase of force to overcome the resistance of the spring. In Naylor's "compensating safety-valve," Fig. 14, the spring is attached to the point *A* of the bent lever, so that as the lift of the valve increases, the lever arm of the resistance diminishes, which has the same effect as increasing the pressure on the valve.

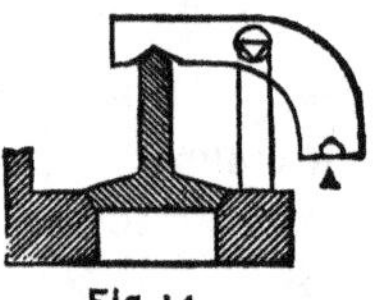

Fig 14

An arrangement for giving a large opening with a comparatively small lift, by having a double-seated valve, has been proposed. A sketch of the plan is shown in Fig. 15, *A* and *B* being the

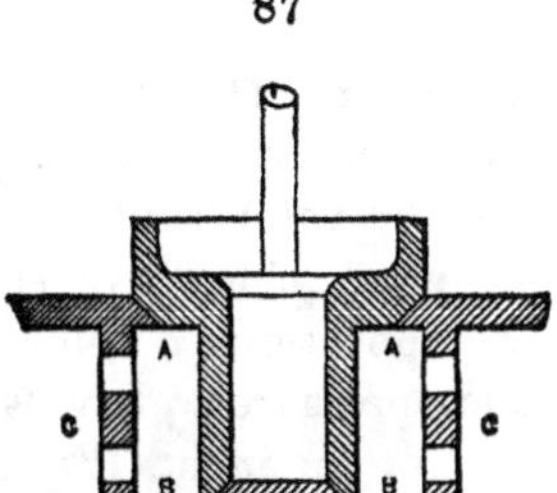

Fig 15

two seats. It is found difficult, however, to keep such a valve tight.

Mr. Thomas Adams, of Scotland, constructs a valve with a projecting lip, *A*, Fig. 16. By applying steam gauges at the points shown on the sketch, he observed that they gave the indicated

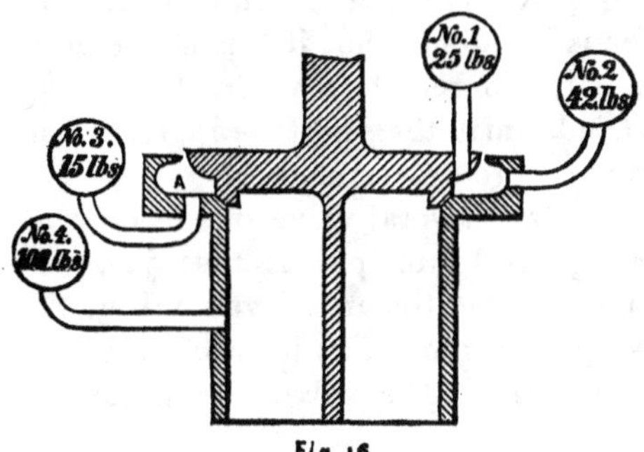

Fig 16

readings at the time the valve opened, and that then the pressures began to lower, around the valve first, and lastly in the boiler, when the valve suddenly closed. He proportions a 3-inch valve for 100 pounds pressure to have a lift of five-sixteenths of an inch. Finding by experiment that the spring used requires a load of 138 pounds to compress it this amount, he makes the area of the lip sufficient to produce this load when acted upon by steam having a pressure of 25 pounds per square inch. All good safety-valves must be proportioned somewhat after this manner. It is better to make the direct experiment with steam gauges for any particular case, as Mr. Adams has done ; but if this is not convenient, different forms of valves must be tried, until the right proportions of the parts are determined.

There are several valves made on the same general principle as that just described. The Richardson valve, Fig. 17, is very well known, and largely used in this country. The valve has a curved lip, *A*, and there is a recess, *B*, surrounding the seat. It is claimed by the inventor that the recoil of the escaping steam produced by this arrangement gives a higher lift to the valve. Some experiments were made with this valve by or-

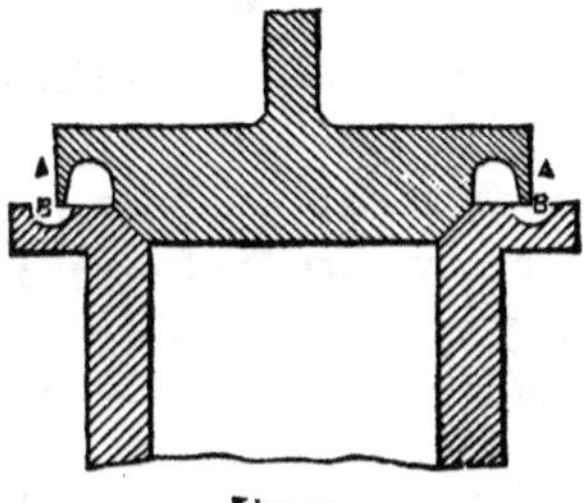

Fig. 17

der of the Navy Department, in 1873. The boiler to which it was attached was first fitted with a common valve 4 inches in diameter, which let the steam pressure increase in 4 minutes 5 pounds beyond the point for which it was set, when the furnace doors were closed, and a 3-inch common valve allowed the pressure to increase 9 pounds in 6 minutes, under the same circumstances. A Richardson valve, 3 inches in diameter, being attached, the pressure could not be increased beyond the point for which the valve was set, and it closed promptly on a slight reduction of the steam pressure. The common safety-valve being set to open at different pressures gave the same amount of lift in each case, as far as could be observed, while the Richardson valve gave an increased lift for each increase of pressure at which it was set.

A competitive trial of safety-valves was recently conducted in England, for a prize of $500 offered by the Editor of the *Nautical Magazine.* A sketch of the Rochford valve, to which the prize was awarded, is shown in Fig. 18. It will be seen that it has a projecting lip, *A*, and the spring is enclosed in a case, *B*, which can be filled with oil if desired, for the preservation of the spring. In the trial the valve was attached to a Lancashire boiler, 33 feet long, 7 feet 2 inches in diameter, flues 2 feet 8 inches in diameter, grate surface 30 square feet; being set to blow off at 60 pounds, the pressure could not be increased beyond

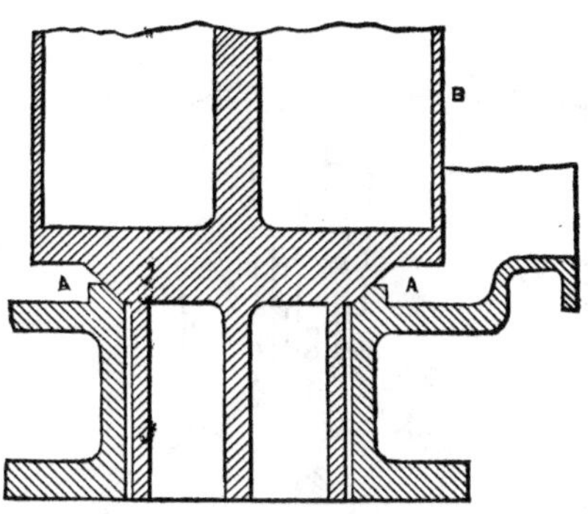

Fig 18

$63\frac{1}{2}$ pounds by the most severe firing, and when it fell to 60 pounds, the valve closed promptly.

Instead of having a projecting lip on the valve, a piston is sometimes attached to it, which is exposed to the action of steam at rest. The Rochow valve, Fig. 19, furnishes a good illustration of this principle. A piston, *A*, is attached by a stem to the valve, and the tube in which it works is extended into the boiler by a pipe, *C*, so as to be acted upon by quiescent steam. A ring, *B*, nearly closes the orifice by which steam is admitted to the

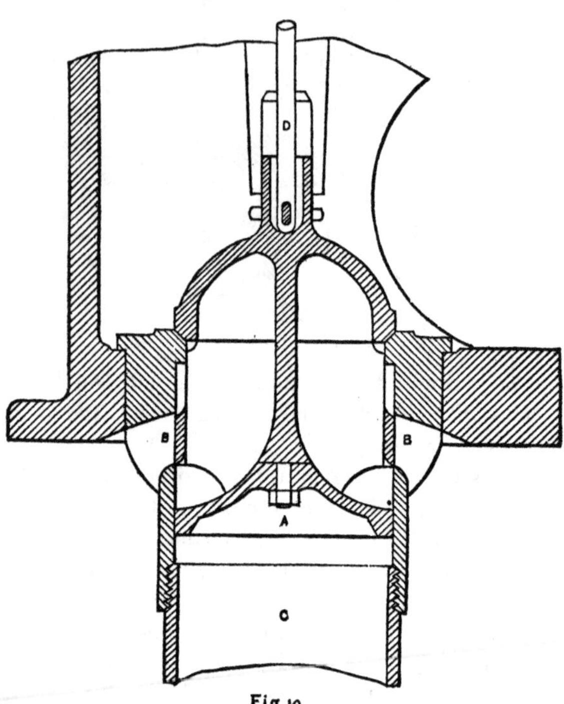

Fig 19

valve proper, when it is seated ; and as the valve rises, the ring opens the orifice more and more. From this description it will be evident that when the valve rises a little, the pressure above the piston will be reduced, and consequently there will be an unbalanced pressure on the under side of the piston to raise the valve higher, and by giving a proper proportion to this piston any required lift can be obtained. In the form represented in the sketch, the valve is arranged to be loaded with a weight, and there is a stop at *D* to check the valve when it has lifted high enough. The piston and ring do not work steam tight, but are fitted loosely, so that they can move without friction.

In these illustrations, while only a few of the most prominent valves have been noticed, it is believed that no important principle of construction has been overlooked. It is a matter of regret that more information cannot be given on this part of the subject, but, as before remarked, the record of experiments from which general rules can be deduced is very slight. Enough has been said, however, to show that there are good safety valves in the market.

There is one experiment that everyone who uses a safety valve can make, to see whether he has a valve that is proportioned to give the proper lift. Let him secure a cord to the lever or stem of the valve, so that it can be opened by hand if necessary. (Indeed, it may be said, in passing, that some convenient arrangement should always be fitted for opening the valve by hand, and it should be used at least once a day, to keep the valve in working order.) Then, by shutting off steam from the engine or wherever else it is used, and making up a good fire in the boiler, he can determine in a very short time whether or not he has a *safety* valve; and if it will not relieve the boiler automatically it will be easy to give a larger opening by hand, so that the experiment will not be attended with danger. This simple experiment is earnestly recommended to every steam user, for with a good safety-valve in working order the chances of a disastrous boiler explosion are greatly diminished.

V. The Relative Merits of Valves Loaded with Weights and Springs.

The best manner of loading a safety-valve has been the subject of animated discussion among engineers. The opinion of the majority can be summed up as follows :

Safety-valves for the boilers of locomotives and steamers, and in all cases in which they will be subjected to oscillations and jars, should be loaded with springs. For stationary boilers, either weights or springs can be used at pleasure. In employing a spring it is generally considered best to arrange it so that it shall be compressed, rather than extended, when the valve is raised.

APPENDIX.

The demonstrations of the rules given in a former part of this book, for finding the opening due to a given lift of a valve, when it does not rise clear of the seat, and for determining the area of opening necessary to discharge a given weight of steam of known pressure, are appended.

a.—The area of opening of a valve with a beveled seat, for a lift $A\ E = l$, Fig. 20, is the area of the frustum of a cone whose upper base $A\ B = D$, has the same diameter as the valve, and whose lower base, $C\ D$, measured from points, C, D, where perpendiculars from A and B intersect the seat, is equal to

$$D + C\,G + H\,D = D + 2\ C\,G$$

the slant height being $A\ C$ or $B\ D$. The

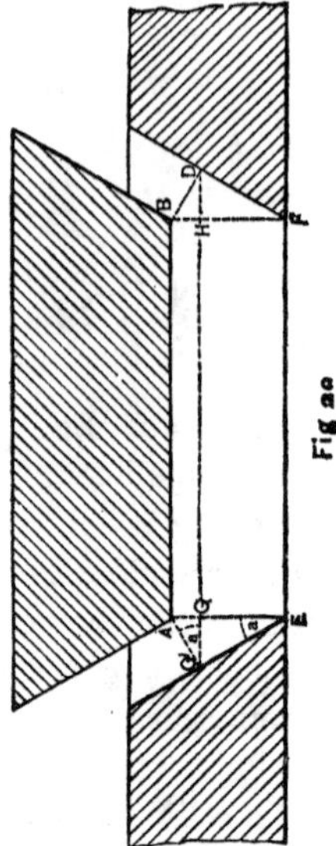

area of the surface is equal to the half sum of the circumferences of the upper and lower bases multiplied by the slant height, or the area of opening due to the

$$\text{lift } l = \frac{3.1416\ (2\ D + C\ G)}{2} \times A\ C.$$

The angle of bevel, $A\ E\ C = a$, and

the angle $A\ C\ G$ is also equal a. From trigonometry, in the triangle $A\ E\ C$,

$$\text{Sine } a = \frac{A\ C}{A\ E} = \frac{A\ C}{l}$$

$$A\ C = l \times \text{sine } a.$$

In the triangle $A\ C\ G$,

$$\text{Cosine } a = \frac{C\ G}{A\ C} \cdot = \frac{C\ G}{l \times \text{sine } a.}$$

$$C\ G = l \times \text{sine } a \times \text{cosine } a$$

Substituting these values in the previous equation :

Area of opening—

$$= \frac{3.1416\,(2\ D \times 2\ l \times \text{sine } a \times \text{cosine } a)}{2} \times l \times \text{sine } a$$

$$= 3.1416\,[D \times l \times \text{sine } a + l^2 \times (\text{sine } a)^2 \times \text{cosine } a]$$

If $a = 45°$,

$$\text{Sine } a = 0.707, \text{ cosine } a = 0.707$$

$$(\text{Sine } a)^2 = 0.5$$

$$(\text{Sine } a)^2 \times \cos a = 0.3535$$

Whence, area of opening—

$$= 3.1416\,(D \times l \times 0.707 + l^2 \times 0.3535)$$

$$= 2.22 \times D \times l + 1.11\ l^2$$

If $a = 30°$,

Sine $a = 0.5$, cosine $a = 0.866$
$(\text{Sine } a)^2 = 0.25$
$(\text{Sine } a)^2 \times \cos a = 0.2165$.

Area of opening—

$= 3.1416 (D \times l \times 0.5 + l^2 \times 0.2165$
$= 1.57 \times D \times l + 0.68 \times l^2$

b.—Let

A = area of opening required to discharge W pounds of steam at the pressure P per hour. The quantity discharged by this opening per second will be, approximately,

$$\frac{P \times A}{70}$$

Hence the quantity discharged per hour,

$$W = \frac{P \times A \times 3600}{70}$$

$= 51.43\ P \times A$, nearly,

And—

$$A = \frac{W}{51.43 \times P}$$

**** Any book in this Catalogue sent free by mail on receipt of price.*

VALUABLE
SCIENTIFIC BOOKS,

PUBLISHED BY

D. VAN NOSTRAND,

23 MURRAY STREET AND 27 WARREN STREET,

NEW YORK.

FRANCIS. Lowell Hydraulic Experiments, being a selection from Experiments on Hydraulic Motors, on the Flow of Water over Weirs, in Open Canals of Uniform Rectangular Section, and through submerged Orifices and diverging Tubes. Made at Lowell, Massachusetts. By James B. Francis, C. E. 2d edition, revised and enlarged, with many new experiments, and illustrated with twenty-three copperplate engravings. 1 vol. 4to, cloth......................$15 00

ROEBLING (J. A.) Long and Short Span Railway Bridges. By John A. Roebling, C. E. Illustrated with large copperplate engravings of plans and views. Imperial folio, cloth............................ 25 00

CLARKE (T. C.) Description of the Iron Railway Bridge over the Mississippi River, at Quincy, Illinois. Thomas Curtis Clarke, Chief Engineer. Illustrated with 21 lithographed plans. 1 vol. 4to, cloth...................................... 7 50

TUNNER (P.) A Treatise on Roll-Turning for the Manufacture of Iron. By Peter Tunner. Translated and adapted by John B. Pearse, of the Penn-

,ylvania Steel Works, with numerous engravings wood cuts and folio atlas of plates.................$10 00

ISHERWOOD (B. F.) Engineering Precedents for Steam Machinery. Arranged in the most practical and useful manner for Engineers. By B. F. Isherwood, Civil Engineer, U. S. Navy. With Illustrations. Two volumes in one. 8vo, cloth........... $2 50

BAUERMAN. Treatise on the Metallurgy of Iron, containing outlines of the History of Iron Manufacture, methods of Assay, and analysis of Iron Ores, processes of manufacture of Iron and Steel, etc., etc. By H. Bauerman. First American edition. Revised and enlarged, with an Appendix on the Martin Process for making Steel, from the report of Abram S. Hewitt. Illustrated with numerous wood engravings. 12mo, cloth...................................... 2 00

CAMPIN on the Construction of Iron Roofs. By Francis Campin. 8vo, with plates, cloth........... 2 00

COLLINS. The Private Book of Useful Alloys and Memoranda for Goldsmiths, Jewellers, &c. By James E. Collins. 18mo, cloth.................... 75

CIPHER AND SECRET LETTER AND TELEGRAPHIC CODE, with Hogg's Improvements. The most perfect secret code ever invented or discovered. Impossible to read without the key. By C. S. Larrabee. 18mo, cloth..................... 1 00

COLBURN. The Gas Works of London. By Zerah Colburn, C. E. 1 vol. 12mo, boards............... 60

CRAIG (B. F.) Weights and Measures. An account of the Decimal System, with Tables of Conversion for Commercial and Scientific Uses. By B. F. Craig, M.D. 1 vol. square 32mo, limp cloth.... 50

NUGENT. Treatise on Optics; or, Light and Sight, theoretically and practically treated; with the application to Fine Art and Industrial Pursuits. By E. Nugent. With one hundred and three illustrations. 12mo, cloth....................................... 2 00

FREE HAND DRAWING. A Gnide to Ornamental Figure and Landscape Drawing. By an Art Student. 18mo, boards,.............................. 50

HOWARD. Earthwork Mensuration on the Basis of the Prismoidal Formulae. Containing simple and labor-saving method of obtaining Prismoidal contents directly from End Areas. Illustrated by Examples, and accompanied by Plain Rules for Practical Uses. By Conway R. Howard, C. E., Richmond, Va. Illustrated, 8vo, cloth.......................... 1 50

GRUNER. The Manufacture of Steel. By M. L. Gruner. Translated from the French, by Lenox Smith, with an appendix on the Bessamer process in the United States, by the translator. Illustrated by Lithographed drawings and wood cuts. 8vo, cloth.. 3 50

AUCHINCLOSS. Link and Valve Motions Simplified. Illustrated with 37 wood-cuts, and 21 lithographic plates, together with a Travel Scale, and numerous useful Tables. By W. S. Auchincloss. 8vo, cloth.. 3 00

VAN BUREN. Investigations of Formulas, for the strength of the Iron parts of Steam Machinery. By J. D. Van Buren, Jr., C. E. Illustrated, 8vo, cloth. 2 00

JOYNSON. Designing and Construction of Machine Gearing. Illustrated, 8vo, cloth.................... 2 00

GILLMORE. Coignet Beton and other Artificial Stone. By Q. A. Gillmore, Major U. S. Corps Engineers. 9 plates, views, &c. 8vo, cloth.................... 2 50

SAELTZER. Treattse on Acoustics in connection with Ventilation. By Alexander Saeltzer, Architect. 12mo, cloth.................................... 2 00

THE EARTH'S CRUST. A handy Outline of Geology. By David Page. Illustrated, 18mo, cloth.... 75

DICTIONARY of Manufactures, Mining, Machinery, and the Industrial Arts. By George Dodd. 12mo, cloth.. 2 00

BOW. A Treatise on Bracing, with its application to Bridges and other Structures of Wood or Iron. By Robert Henry Bow, C. E. 156 illustrations, 8vo, cloth.. 1 50

GILLMORE (Gen. Q. A.) Treatise on Limes, Hydraulic Cements, and Mortars. Papers on Practical Engineering, U. S. Engineer Department, No. 9, containing Reports of numerous Experiments conducted in New York City, during the years 1858 to 1861, inclusive. By Q. A. Gillmore, Bvt. Maj.-Gen., U. S. A., Major, Corps of Engineers. With numerous illustrations. 1 vol, 8vo, cloth............... $4 00

HARRISON. The Mechanic's Tool Book, with Practical Rules and Suggestions for Use of Machinists, Iron Workers, and others. By W. B. Harrison, associate editor of the "American Artisan." Illustrated with 44 engravings. 12mo, cloth............ 1 50

HENRICI (Olaus). Skeleton Structures, especially in their application to the Building of Steel and Iron Bridges. By Olaus Henrici. With folding plates and diagrams. 1 vol. 8vo, cloth................... 1 50

HEWSON (Wm.) Principles and Practice of Embanking Lands from River Floods, as applied to the Levees of the Mississippi. By William Hewson, Civil Engineer. 1 vol. 8vo, cloth...................... 2 00

HOLLEY (A. L.) Railway Practice. American and European Railway Practice, in the economical Generation of Steam, including the Materials and Construction of Coal-burning Boilers, Combustion, the Variable Blast, Vaporization, Circulation, Superheating, Supplying and Heating Feed-water, etc., and the Adaptation of Wood and Coke-burning Engines to Coal-burning; and in Permanent Way, including Road-bed, Sleepers, Rails, Joint-fastenings, Street Railways, etc., etc. By Alexander L. Holley, B. P. With 77 lithographed plates. 1 vol. folio, cloth.... 12 00

KING (W. H.) Lessons and Practical Notes on Steam, the Steam Engine, Propellers, etc., etc., for Young Marine Engineers, Students, and others. By the late W. H. King, U. S. Navy. Revised by Chief Engineer J. W. King, U. S. Navy. Twelfth edition, enlarged. 8vo, cloth. 2 00

MINIFIE (Wm.) Mechanical Drawing. A Text-Book of Geometrical Drawing for the use of Mechanics and Schools, in which the Definitions and Rules of Geometry are familiarly explained; the Practical Problems are arranged, from the most simple to the more complex, and in their description technicalities are avoided as much as possible. With illustrations for Drawing Plans, Sections, and Elevations of Railways and Machinery; an Introduction to Isometrical Drawing, and an Essay on Linear Perspective and Shadows. Illustrated with over 200 diagrams engraved on steel. By Wm. Minifie, Architect. Seventh edition. With an Appendix on the Theory and Application of Colors. 1 vol. 8vo, cloth........... $4 00

"It is the best work on Drawing that we have ever seen, and is especially a text-book of Geometrical Drawing for the use of Mechanics and Schools. No young Mechanic, such as a Machinists, Engineer, Cabinet-maker, Millwright, or Carpenter, should be without it."—*Scientific American.*

—— Geometrical Drawing. Abridged from the octavo edition, for the use of Schools. Illustrated with 48 steel plates. Fifth edition. 1 vol. 12mo, cloth.... 2 00

STILLMAN (Paul.) Steam Engine Indicator, and the Improved Manometer Steam and Vacuum Gauges—their Utility and Application. By Paul Stillman. New edition. 1 vol. 12mo, flexible cloth........... 1 00

SWEET (S. H.) Special Report on Coal; showing its Distribution, Classification, and cost delivered over different routes to various points in the State of New York, and the principal cities on the Atlantic Coast. By S. H. Sweet. With maps, 1 vol. 8vo, cloth..... 3 00

WALKER (W. H.) Screw Propulsion. Notes on Screw Propulsion: its Rise and History. By Capt. W. H. Walker, U. S. Navy. 1 vol. 8vo, cloth..... 75

WARD (J. H.) Steam for the Million. A popular Treatise on Steam and its Application to the Useful Arts, especially to Navigation. By J. H. Ward, Commander U. S. Navy. New and revised edition. 1 vol. 8vo, cloth.................................. 1 00

WEISBACH (Julius). Principles of the Mechanics of Machinery and Engineering. By Dr. Julius Weisbach, of Freiburg. Translated from the last German edition. Vol. I., 8vo, cloth.. .... 10 00

DIEDRICH. The Theory of Strains, a Compendium for the calculation and construction of Bridges, Roofs, and Cranes, with the application of Trigonometrical Notes, containing the most comprehensive information in regard to the Resulting strains for a permanent Load, as also for a combined (Permanent and Rolling) Load. In two sections, adadted to the requirements of the present time. By John Diedrich, C. E. Illustrated by numerous plates and diagrams. 8vo, cloth.................................... 5 00

WILLIAMSON (R. S.) On the use of the Barometer on Surveys and Reconnoissances. Part I. Meteorology in its Connection with Hypsometry. Part II. Barometric Hypsometry. By R. S. Wiliamson, Bvt Lieut.-Col. U. S. A., Major Corps of Engineers. With Illustrative Tables and Engravings. Paper No. 15, Professional Papers, Corps of Engineers. 1 vol. 4to, cloth.................................. 15 00

POOK (S. M.) Method of Comparing the Lines and Draughting Vessels Propelled by Sail or Steam. Including a chapter on Laying off on the Mould-Loft Floor. By Samuel M. Pook, Naval Constructor. 1 vol. 8vo, with illustrations, cloth............ 5 00

ALEXANDER (J. H.) Universal Dictionary of Weights and Measures, Ancient and Modern, reduced to the standards of the United States of America. By J. H. Alexander. New edition, enlarged. 1 vol. 8vo, cloth.................................. 3 50

WANKLYN. A Practical Treatise on the Examination of Milk, and its Derivatives, Cream, Butter and Cheese. By J. Alfred Wanklyn, M. R. C. S., 12mo cloth.. 1 00

RICHARDS' INDICATOR. A Treatise on the Richards Steam Engine Indicator, with an Appendix by F. W. Bacon, M. E. 18mo, flexible, cloth.......... 1 00

POPE. Modern Practice of the Electric Telegraph. A Hand Book for Electricians and operators. By Frank L. Pope. Eighth edition, revised and enlarged, and fully illlustrated. 8vo, cloth........................ $2.00

"There is no other work of this kind in the English language that contains in so small a compass so much practical information in the application of galvanic electricity to telegraphy. It should be in the hands of every one interested in telegraphy, or the use of Batteries for other purposes."

MORSE. Examination of the Telegraphic Apparatus and the Processes in Telegraphy. By Samuel F. Morse, LL.D., U. S. Commissioner Paris Universal Exposition, 1867. Illustrated, 8vo, cloth.......... $2 00

SABINE. History and Progress of the Electric Telegraph, with descriptions of some of the apparatus. By Robert Sabine, C. E. Second edition, with additions, 12mo, cloth.............................. 1 25

BLAKE. Ceramic Art. A Report on Pottery, Porcelain, Tiles, Terra Cotta and Brick. By W. P. Blake, U. S. Commissioner, Vienna Exhibition, 1873. 8vo, cloth.. 2 00

BENET. Electro-Ballistic Machines, and the Schultz Chronoscope. By Lieut.-Col. S. V. Benet, Captain of Ordnance, U. S. Army. Illustrated, second edition, 4to, cloth.................................. 3 00

MICHAELIS. The Le Boulenge Chronograph, with three Lithograph folding plates of illustrations. By Brevet Captain O. E. Michaelis, First Lieutenant Ordnance Corps, U. S. Army, 4to, cloth.......... 3 00

ENGINEERING FACTS AND FIGURES An Annual Register of Progress in Mechanical Engineering and Construction, for the years 1863, 64, 65, 66, 67, 68. Fully illustrated, 6 vols. 18mo, cloth, $2.50 per vol., each volume sold separately..............

HAMILTON. Useful Information for Railway Men. Compiled by W. G. Hamilton, Engineer. Fifth edition, revised and enlarged, 562 pages Pocket form. Morocco, gilt.................................... 2 00

STUART. The Civil and Military Engineers of America. By Gen. C. B. Stuart. With 9 finely executed portraits of eminent engineers, and illustrated by engravings of some of the most important works constructed in America. 8vo, cloth.................... $5 00

STONEY. The Theory of Strains in Girders and similar structures, with observations on the application of Theory to Practice, and Tables of Strength and other properties of Materials. By Bindon B. Stoney, B. A. New and revised edition, enlarged, with numerous engravings on wood, by Oldham. Royal 8vo, 664 pages. Complete in one volume. 8vo, cloth....... 12 50

SHREVE. A Treatise on the Strength of Bridges and Roofs. Comprising the determination of Algebraic formulas for strains in Horizontal, Inclined or Rafter, Triangular, Bowstring, Lenticular and other Trusses, from fixed and moving loads, with practical applications and examples, for the use of Students and Engineers. By Samuel H. Shreve, A. M., Civil Engineer. 87 wood cut illustrations. 8vo, cloth............... 5 00

MERRILL. Iron Truss Bridges for Railroads. The method of calculating strains in Trusses, with a careful comparison of the most prominent Trusses, in reference to economy in combination, etc., etc. By Brevet Col. William E. Merrill, U. S. A., Major Corps of Engineers, with nine lithographed plates of Illustrations. 4to, cloth.......................... 5 00

WHIPPLE. An Elementary and Practical Treatise on Bridge Building. An enlarged and improved edition of the author's original work. By S. Whipple, C. E., inventor of the Whipple Bridges, &c. Illustrated 8vo, cloth.. 4 00

THE KANSAS CITY BRIDGE. With an account of the Regimen of the Missouri River, and a description of the methods used for Founding in that River. By O. Chanute, Chief Engineer, and George Morrison, Assistant Engineer. Illustrated with five lithographic views and twelve plates of plans. 4to, cloth, 6 00

www.ingramcontent.com/pod-product-compliance
Lightning Source LLC
LaVergne TN
LVHW021429110826
845150LV00007B/2152

9781425507107